WEST CHICAGO PUBLIC LIBRARY DISTRICT
3 6653 00236 5033
P9-CML-117

REVOLUTIONARY DISCOVERIES OF SCIENTIFIC PIONEERS™

LAWS OF MOTION AND ISAAC NEWTON

FRED BORTZ

Published in 2014 by The Rosen Publishing Group, Inc.
29 East 21st Street, New York, NY 10010

First Edition

Library of Congress Cataloging-in-Publication Data

Bortz, Fred, 1944–
Laws of motion and Isaac Newton/Fred Bortz.—First edition.
pages cm.—(revolutionary discoveries of scientific pioneers)
Audience: Grades 7-12.
Includes bibliographical references and index.
ISBN 978-1-4777-1808-7 (library binding)
1. Newton, Isaac, 1642–1727—Juvenile literature. 2. Physicists—Great Britain—Biography—Juvenile literature. 3. Motion—Juvenile literature. I. Title.
QC16.N7B547 2014
530.092—dc23
[B]

2013011618

Manufactured in the United States of America

CPSIA Compliance Information: Batch #W14YA: For further information, contact Rosen Publishing, New York, New York, at 1-800-237-9932.

A portion of the material in this book has been derived from *Newton and the Three Laws of Motion* by Nicholas Croce.

CONTENTS

INTRODUCTION

"If I have seen further, it is by standing on the shoulders of giants." If there were a single sentence to summarize the life of Sir Isaac Newton (1642–1727), it would be that quotation from a February 5, 1675, letter to his one of his greatest scientific rivals, Robert Hooke (1635–1703). Though most of Newton's greatest work had yet to unfold when he wrote that letter, his brilliance and his difficult personality were already apparent to the scholars of his time. That combination of mind and personality was almost certain to inspire professional jealousies, and Newton's career was marked by disputes that were almost as great as his breakthroughs in optics (the study of light), mathematics, and mechanics (the study of force and motion).

The giants whom Newton referred to were his predecessors in science and what was then called "natural philosophy." They included the scholars who struggled to make sense of the motion of the sun, moon, planets, and heavenly bodies. Newton stood on their shoulders to discover the scientific principles—his laws of universal gravitation and motion—that explained the mathematical forms and patterns that others had discovered in the motion of those bodies.

His explanation relied on a new branch of mathematics that we now call calculus, invented at about the same time by both Newton and another rival,

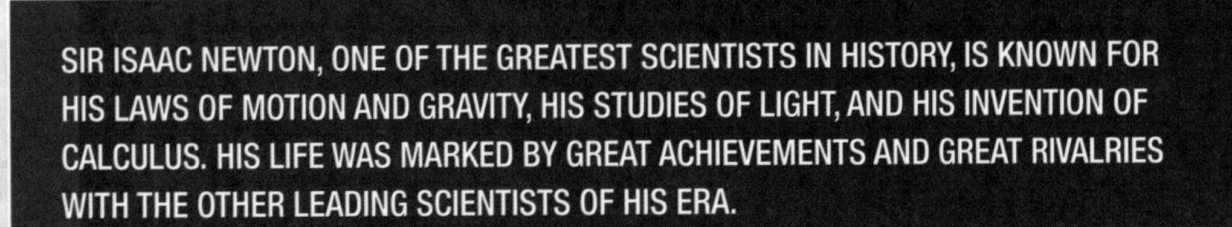

SIR ISAAC NEWTON, ONE OF THE GREATEST SCIENTISTS IN HISTORY, IS KNOWN FOR HIS LAWS OF MOTION AND GRAVITY, HIS STUDIES OF LIGHT, AND HIS INVENTION OF CALCULUS. HIS LIFE WAS MARKED BY GREAT ACHIEVEMENTS AND GREAT RIVALRIES WITH THE OTHER LEADING SCIENTISTS OF HIS ERA.

Gottfried Leibniz (1646–1716). Newton also discovered and described many important principles of optics. That included asking what turned out to be one of science's greatest questions: Is light made of particles or waves? Though Hooke and another rival, Christiaan Huygens (1629–1695), disputed his answer, it eventually led to one of the greatest breakthroughs of twentieth-century science.

Today, Newton is regarded as one of the greatest giants on whose shoulders generations of scientists have stood ever since. One of those, Albert Einstein (1879–1955), discovered the theory of relativity. Einstein's theory restated Newton's laws for motion at nearly the speed of light. It also led to new understandings and new questions about gravity, the nature of matter and energy, and properties of the universe itself.

At about the same time as Einstein proposed his theory, other scientists were making discoveries that led to an unexpected conclusion about light. Light is neither a wave nor a stream of particles but a little bit of both—an answer that Newton and his rivals could never have anticipated. That discovery opened a new branch of physics known as quantum mechanics, which has produced much of the remarkable technology that has transformed the modern world.

Modern science is the result of giants on the shoulders of giants on the shoulders of other giants. And one of the greatest giants of all is Isaac Newton.

CHAPTER ONE

A REMARKABLE LIFE

Scientific genius and his rivalries are only part of the Isaac Newton story. Often insecure and anxious, Newton experienced emotional turmoil and even mental breakdowns during his long career. Despite those difficulties, or perhaps because of how he responded to them, he managed to transform scientific thought for all time.

Newton's remarkable life was difficult from its beginning in the small village of Woolsthorpe in Lincolnshire, England. Born on Christmas Day of 1642, he was so sickly that he was not even expected to survive his first day of life, let alone live for more than eighty-four years. His father had died three months before, so from birth, the young Isaac had only his mother to care for him. Within two years, she married the wealthy

NEWTON WAS BORN TO A WIDOWED MOTHER IN THIS HOME IN THE SMALL VILLAGE OF WOOLSTHORPE, LINCOLNSHIRE, ENGLAND. HE WAS SO SICKLY AT BIRTH THAT HE WAS NOT EXPECTED TO SURVIVE FOR EVEN A DAY. FORTUNATELY FOR THE WORLD OF SCIENCE, HE LIVED FOR 84 YEARS.

minister Barnabas Smith and left Isaac to be raised by his grandmother. Isaac's mother did not return to him until Smith died nine years later in 1653.

NEWTON'S EDUCATION

Newton left home to attend Trinity College in Cambridge, England, in June 1661. There he studied with professors who were shaping a period of history we now call the Scientific Revolution. Great advances in knowledge were taking place, and the young Newton studied them intently. He read the works of Aristotle, the ancient Greek philosopher and scientist. At the same time, Newton also began to study the works of new thinkers like French mathematician and scientist René Descartes (1596–1650). He also learned about German astronomer Johannes Kepler (1571–1630) and the three laws he outlined to describe planetary motion. He read the writings of Italian scientist Galileo Galilei (1564–1642), who was the first person to look into space with a telescope.

Newton was so inspired by what he read that in 1664, at the age of only twenty-one, he published his own philosophy. He began the work, *Quaestiones Quaedam Philosophicae* (*Certain Philosophical Questions*), with the now-famous line "*Amicus Plato amicus Aristoteles magis amica veritas.*" Translated from Latin, this phrase reads: "Plato is my friend, Aristotle is my friend, but my best friend is truth." What Newton meant by these words was that though he respected the esteemed minds of the past, he wanted to break new ground and make discoveries of his own, even if those discoveries contradicted those great thinkers.

With this newfound ambition, Newton left Trinity in April 1665 with a bachelor's degree. He would have continued with his education, but then disaster struck.

THE BLACK DEATH

Almost immediately after Newton graduated, the deadly bubonic plague began to spread rapidly through Europe, killing thousands in its path. Called the Black Death because of the discoloration it formed on its victims, the bubonic plague led

AT TRINITY COLLEGE OF CAMBRIDGE UNIVERSITY, NEWTON'S STUDIES INCLUDED THE CLASSIC TEXTS OF ANCIENT SCHOLARS, SUCH AS ARISTOTLE, SHOWN IN THIS FAMOUS PAINTING BY REMBRANDT CONTEMPLATING THE BUST OF HOMER. NEWTON HELD THE WORKS OF THE ANCIENTS IN HIGH ESTEEM, BUT HE ALSO KNEW THAT NEW KNOWLEDGE COULD CONTRADICT WHAT THEY BELIEVED TO BE TRUE.

English officials to close the doors of most public institutions, including Trinity College, and to enforce rules to confine people to their homes, limiting interaction between people in hopes of curbing the spread of the disease.

For the next two years, Newton rarely left his house. This, however, gave him plenty of time to study and think. Unmarried and having few, if any, friends, Newton devoured books, especially those concerning sciences. Newton studied the various branches of mathematics, including algebra and geometry. He mastered many difficult concepts in little more than a year.

Additionally, while the plague paralyzed England, Newton began thinking about motion of both everyday objects and the bodies of the universe. He discovered mathematical ways to describe the way objects interacted and moved. He realized that the same mathematical formulas describe the motion of both objects on Earth and heavenly bodies. Today we call the study of the motion of objects mechanics. It is a major branch of the field of physics.

As Newton studied the movement of the moon and planets, he realized that the same force of gravity that made objects fall on Earth influenced the motion of heavenly bodies as well. This was the foundation for his theory of universal gravitation, but it would be many years before Newton would be able to express that law in mathematical form.

THE NATURE AND PROPERTIES OF LIGHT

Newton's interests went beyond mathematics and mechanics. He also wanted to understand the properties of light. What causes colors? Why is the sky blue? Why can we see our reflection in a mirror? These were only a few of the questions Newton pursued. Not only was his work in optics, the study of light, important in its own right, but his discoveries about the behavior of light also changed the way people thought about other branches of science.

In 1666, Newton began to study color. Like many great scientific discoveries, what he learned began simply by following his own curiosity. What was light made of and how did it work? Newton began with a prism—a triangular piece of glass—that he had bought at a local fair. He placed it near a window where a ray of sunlight shone through it. He noticed that the prism refracted, or bent, the sunlight and separated it into a band of colors like those of a rainbow. Each color refracted a little bit differently than the others. Another identical prism could reverse the process and combine the colors into white light.

From his experiments, Newton concluded that natural light is made up of many different colors. Each color has unique properties, which cause the differences in refraction and which the human eye can distinguish. Newton's understanding of light allowed him to build the first reflecting telescope. Reflecting telescopes are different from the refracting telescopes that seventeenth-century astronomers were using to study the stars and planets.

Those refracting telescopes used pieces of curved glass called lenses to bend light and create images. Newton figured out how to do the same thing with curved mirrors. Because his mirrors were larger than lenses, they could collect more light, allowing observers to see objects that were dimmer or farther away. They could also be made to focus light a shorter distance away, so that a reflector could produce its images in a tube only one-tenth the length needed for a refractor of similar magnifying power.

NEWTON'S DIFFICULT PERSONALITY LED TO PROFESSIONAL RIVALRIES WITH MANY OF THE OTHER GREAT SCIENTISTS OF HIS TIME. HE AND THE GREAT DUTCH MATHEMATICIAN AND ASTRONOMER CHRISTIAAN HUYGENS, SHOWN HERE BUILDING THE FIRST PENDULUM CLOCK, CLASHED OVER THEIR IDEAS ABOUT THE NATURE OF LIGHT.

From his experiments, Newton concluded that light was a stream of tiny particles that he called corpuscles. He concluded that corpuscles of different colors behaved differently as they passed through transparent materials such as glass or water. Robert Hooke and Christiaan Huygens disagreed. They considered light to be vibrations or waves. Hooke considered himself a master in optics, and his criticisms infuriated Newton. When Newton published his paper *An Hypothesis Explaining the Properties of Light* in 1675, Hooke claimed that much of its content was stolen from him. This first accusation of theft by another scientist shook Newton to his core. By 1678, Newton had a nervous breakdown.

NEWTON'S MASTERPIECE, THE *PRINCIPIA*

By 1679 Newton had once again focused his attention on planetary orbits—the way in which planets move around the sun. By this time, Newton had outlined his famous three laws of motion as well as the mathematical form of his theory of universal gravitation. From these ideas emerged Newton's *Philosophiae Naturalis Principia Mathematica* (*Mathematical Principles of Natural Philosophy*), or simply the *Principia*, which is considered perhaps the greatest masterpiece of science every written.

Begun many years earlier as a manuscript titled *De Motu Corporum in Gyrum* (*On the Motion of*

Revolving Bodies), the *Principia* applied the laws of motion and the theory of universal gravitation to the orbital motion of the planets around the sun and the moon around Earth. The idea can be traced to 1666, when Newton was trying to understand the orbit of the moon. He watched an apple fall from a tree (contrary to legend, the apple did not hit him on the head) and made the connection: the force that pulled the apple down to the ground could extend out into space, even as far as the moon. That force could be responsible for keeping the moon in its orbit around Earth instead of flying off into space. He named that force *gravitas* (gravity) after the Latin word for "heaviness." The *Principia* states that the law of gravity applies to the sun and the planets in exactly the same way as it applies to the moon and Earth.

RIVALRY AND RECOGNITION

After the *Principia* was published, Robert Hooke emerged once again to accuse Newton of thievery. Hooke's claim came from the fact that he had mentioned the idea of gravity to Newton some years before the publication of the *Principia*. In the early editions of the book, Newton had acknowledged ideas from Hooke and several others. But once Hooke accused Newton of stealing his ideas, Newton lashed out and deleted all references to Hooke in a later edition. That further fueled the ongoing rivalry between the two scientists.

The *Principia* raised Newton to international fame. Scientists all over the world, who fervently followed Newton, named themselves Newtonians. Newton was also becoming popular with women, a new and probably disconcerting experience for this formerly reclusive man whose only prior contacts with women were his mother and niece. He was also plagued by the ongoing intense rivalries. Angry and bitter, Newton was nearly driven mad trying to protect his ideas; he felt that rival scientists were trying to profit from them. As a result, Newton left science altogether at the height of his career. In 1696, seventeen years after publishing the *Principia*, he took the job of warden of the British mint, an endeavor that seemed to have nothing to do with his scientific talents.

For the remaining years of his life, Newton gained numerous awards for his lifetime achievements. In 1703, he was elected president of the Royal Society, a prestigious British academy of science, and in 1705, Queen Anne of England knighted him, which was the first time in history a scientist had been given the honor.

Newton continued to revise his work and publish updated editions of his books as well. In 1706 he published a Latin edition of *Opticks*, which outlined his study of light; a second English edition was issued in 1718. However, Newton clearly had passed his glory days. He presided at the meetings of the Royal Society but was often found dozing off.

Newton lived a life remarkable for both its successes and its difficulties. When he died in London on

HOW NEWTON FOILED COUNTERFEITERS

There was very little that was scientific about Newton's post as warden of the British mint. But genius cannot be held down long. Soon after arriving at the mint, Newton developed an expertise in one crucial area—catching counterfeiters. He aided officials at the mint by enforcing strict disciplinary measures over the mint's employees. Newton's work sent many counterfeiters to the gallows to be hanged, a punishment typical of the time for such an offense. Because of these harsh disciplinary practices, Newton had earned the reputation as one of the most feared men by the London underworld.

But disciplining would-be criminals was not enough for Newton. He preferred to find a way to stop counterfeiting altogether. Newton invented the technique called milling—applying ridges on the edges of coins to keep counterfeiters from shaving off the edges of coins and using the precious metals to make new ones. Although milling no longer serves any practical use because coins are no longer made of precious metals, it remains on certain coins today for artistic reasons.

March 31, 1727, he was wealthy from the fortune he amassed at the mint. His true legacy was not money but revolutionary breakthroughs in knowledge and technology: his laws of motion and gravity, his invention of calculus, and his work on optics, including the invention of the reflecting telescope.

That legacy continues as the foundation of modern science and as fuel for the discoveries and inventions of today's "high-tech" world.

CHAPTER TWO

THE SCIENTIFIC REVOLUTION BEFORE NEWTON

Newton was a key figure in a period of great change that many historians call the Scientific Revolution. It began in sixteenth-century Europe when Nicolaus Copernicus (1473–1543) proposed an idea that challenged the way people viewed the universe. Instead of having a stationary Earth in the center of everything (a geocentric universe), Copernicus said that the sun was the center of the universe (a heliocentric universe). In his view, Earth was a planet in orbit around the sun, just like the other five known at the time (Mercury, Venus, Mars, Jupiter, and Saturn).

As the seventeenth century began, Johannes Kepler began to measure the changing positions of the other planets in the sky. He recorded their motions in careful detail. From his measurements, he developed a mathematical description of the

Faksimile einer alten Darstellung des Weltgebäudes nach der Vorstellung des Kopernikus
Nach Andreae Cellarii „Harmonia Macrocosmica“ vom Jahre 1660

THIS “PLANISPHERE” SHOWS COPERNICUS’S HELIOCENTRIC (SUN-CENTERED) MODEL OF THE UNIVERSE AS UNDERSTOOD IN THE EARLY 17TH CENTURY. ORBITING THE SUN IN PERFECT CIRCLES ARE THE PLANETS MERCURY, VENUS, EARTH (WITH THE MOON), MARS, JUPITER (WITH FOUR MOONS DISCOVERED BY GALILEO), AND SATURN. JOHANNES KEPLER’S OBSERVATIONS LATER SHOWED THAT THE ORBITS WERE ACTUALLY ELLIPSES, AND, STILL LATER, NEWTON’S LAWS OF MOTION AND GRAVITY PROVIDED A BASIS FOR KEPLER’S CONCLUSIONS.

shapes of the planets’ orbits, the changing speeds of the planets as they followed those paths, and a relationship between the size of each orbit and the time to complete a trip around it.

During the same period, Galileo Galilei turned the newly invented telescope toward the skies and saw things that convinced him, and many other scientists, that Copernicus's heliocentric idea was correct. But until Newton discovered and described the natural laws that governed those heavenly motions, people continued to doubt Earth's place in the universe.

THE POWER OF THE CHURCH

Before the Scientific Revolution, the Roman Catholic Church was highly esteemed and very powerful in European society. It was the source of authoritative knowledge about the world and the heavens. Why is there day and night? Why does the sun rise in the east and set in the west? Why does the moon have phases? These questions all had one simple answer: They were the work of God. As a result, there wasn't much need for scientific questioning.

The day-to-day lifestyle of people had not improved much for centuries. There was not much fertile land on which to grow crops, and where there was the land, it was not used very efficiently. Large unusable forests covered the land, and malaria, a deadly disease, was common. The population was small and grew very slowly, not exceeding 5 million on the island of Great Britain. (Today Great Britain's population approaches 60 million.) The food was not nutritious, and pestilence and famine were common.

When the Scientific Revolution began, great advances in science swept across Europe. Scientist after scientist made landmark discoveries or greatly improved upon a predecessor's work. It was a time of scientific brainstorming in which some of the universe's greatest riddles were answered. With these advances in science, people began to see the world as they had never seen it before. Before, they viewed its motions as orderly but in a mysterious way controlled by the mind of God. Science enabled them to penetrate the mystery. God's mind was partially revealed to them in the form of natural laws.

PUTTING EARTH IN ITS PLACE

Many historians place the beginning of the Scientific Revolution in the year 1543, when a dying Copernicus published his great work, *De Revolutionibus Orbium Coelestium* (*On the Revolutions of Heavenly Spheres*), commonly called *De Revolutionibus*. Before this time, people commonly accepted a geocentric universe. Challenging that theory was difficult since the Roman Catholic Church supported it as official doctrine. The Earth, Copernicus argued, as well as the rest of the planets, revolved around the sun. Copernicus's theory was in contrast to the dominant model of the motion of the universe.

The church's geocentric doctrine followed the ideas of the second-century Egyptian-Greek astronomer Claudius Ptolemaeus, or Ptolemy. Ptolemy's description of the universe was an improvement on the work

SECOND-CENTURY EGYPTIAN-BORN GREEK SCHOLAR PTOLEMY DESCRIBED THE UNIVERSE AS GEOCENTRIC (EARTH-CENTERED) IN A BOOK CALLED THE *ALMAGEST*. EVEN THOUGH ASTRONOMICAL MEASUREMENTS REQUIRED SCHOLARS TO ADD COMPLICATIONS TO PTOLEMY'S DESCRIPTION, THE GEOCENTRIC MODEL WAS NOT CHALLENGED UNTIL COPERNICUS PUBLISHED HIS BOOK *DE REVOLUTIONIBUS ORBIUM COELESTIUM* (*ON THE REVOLUTIONS OF HEAVENLY SPHERES*).

of the great Greek philosopher Aristotle (384–322 BCE), who viewed the universe as a set of interconnected spheres rotating around the Earth. According to Aristotle, motion of heavenly bodies had to be in perfect circles at constant speed, and Earth was at the center of everything.

That requirement led to some mathematical complications, which Ptolemy described in an enormous volume entitled the *Almagest*. In that book, he added three mathematical elements that brought the observed motions of the planets in line with the idea that heavenly objects must move in perfect circles at constant speeds. Instead of having Earth at the exact center of the planet's circular path, he added a point called the equant.

The sun traveled in a circle around Earth, but its speed on that circle was not constant. Instead, the angle it made with the equant changed at a constant rate. The midpoint of the line segment between Earth and the equant was the center of a circle called the deferent, or the main circular path, for each planet. The equant and deferent were still not enough to solve the mismatch between his math and the actual motion, so for each planet, he added a smaller circle, called an epicycle, going around a point on the larger circle. That seemed to solve the problem.

But over time, as astronomers gathered more measurements with better tools, Ptolemy's description needed to be modified. The solution was simple to devise, but it made the motion seem even more complicated. They added epicycles to the epicycles. By Copernicus's time, Ptolemy's description had become much more complicated than originally envisioned. The heliocentric model of *De Revolutionibus* removed the need for the equant, the deferent, and some of the epicycles, though Copernicus believed that heavenly perfection still required circular motions.

By putting Earth in motion around the sun and setting it spinning around a tilted axis to account for seasonal changes and day and night, Copernicus had challenged church doctrine. He also set in motion the changes that became the Scientific Revolution.

THE END OF EPICYCLES

Because of the power of the church, Copernicus's ideas were slow to take hold. In 1589, forty-six years after Copernicus's death, the brilliant young Johannes Kepler entered the University of Tübingen, where his professors, including one of the most respected astronomy teachers in Europe, Michael Maestlin, were still teaching the Ptolemaic model of the universe. Outside of class, however, Maestlin introduced Kepler to Copernicus's heliocentric model.

Kepler quickly realized that the sun-centered view was correct. But over time, he began to think the epicycles made it more complex than necessary. While teaching astronomy at the University of Graz beginning in 1594, and later while working in Prague at the observatory built by the great astronomer Tycho Brahe (1546–1601), Kepler began to look for a simpler, clearer way to describe the motion of the planets—and he found it.

In his 1609 book *Astronomia Nova* (*New Astronomy*), Kepler wrote that the planets move around the sun in particular oval-shaped paths called ellipses, rather than in perfectly circular orbits. The need for epicycles was gone, and Kepler offered this as further proof of the heliocentric model.

Astronomia Nova included two of what became known as Kepler's three laws of planetary motion. The first law is that the

BEFORE THE INVENTION OF THE TELESCOPE, TYCHO BRAHE'S OBSERVATORY HAD THE FINEST INSTRUMENTS IN THE WORLD. BRAHE PROPOSED A MODEL OF PLANETARY MOTION THAT WAS A COMPROMISE BETWEEN PTOLEMY'S AND COPERNICUS'S AND HIRED THE MATHEMATICALLY BRILLIANT JOHANNES KEPLER TO PROVE THAT HE WAS CORRECT. AFTER BRAHE'S DEATH, KEPLER GATHERED ADDITIONAL DATA THAT SUPPORTED THE COPERNICAN VIEW. HE ALSO DEVELOPED THREE MATHEMATICAL LAWS THAT DESCRIBED PLANETARY ORBITS WITH GREAT ACCURACY. NEWTON'S LAWS OF MOTION AND UNIVERSAL GRAVITATION LATER PROVIDED A PHYSICAL EXPLANATION FOR KEPLER'S LAWS.

planets' orbits are ellipses with the sun at one focus. A planet doesn't travel at a constant speed around its orbit, but rather moves faster when it is closer to the sun and slower when it is farthest away. Kepler's second law explains precisely how that speed changes. It is often called the law of equal areas. If you mark the orbit into sections covering the same amount of time, the sections have different shapes. When the planet is closer to the sun, the section is wider and not as long as a section from when the sun is farther away. But the areas of the two sections are exactly the same.

Kepler published his third law in his 1619 book, *Harmonice Mundi* (*Harmony of the World*). It describes the relationship between a planet's distance from the sun and the length of its year, or orbital period. The farther a planet is from the sun, the longer it takes to complete one orbit. Mercury and Venus are closer than Earth, so they complete their orbits in less than a year. Mars, Jupiter, and Saturn are farther out, and their orbital periods are longer. These were the only planets he knew about.

He looked for a mathematical formula that would relate the orbital period of a planet to its distance from the sun. He could easily see that it was not as simple as a direct proportion, which would state that twice the distance corresponds to twice the orbital period. But to a mathematical mind like Kepler's, it was not difficult to discover a different direct proportion that fit the measurements.

CIRCLES, ELLIPSES, AND PLANETARY ORBITS

An ellipse can be viewed as a circle that is stretched according to a particular mathematical rule. A circle can be described as the set of points on a flat surface that are all the same distance from another point called the center. In the diagram, the center of the circle is the point marked C. To get an ellipse from that circle, replace the center point with the two points on the diagram labeled A and B and stretch the circle so that it becomes an oval with its longest distance along the line connecting A and B. A and B are called the foci (plural of focus) of the ellipse.

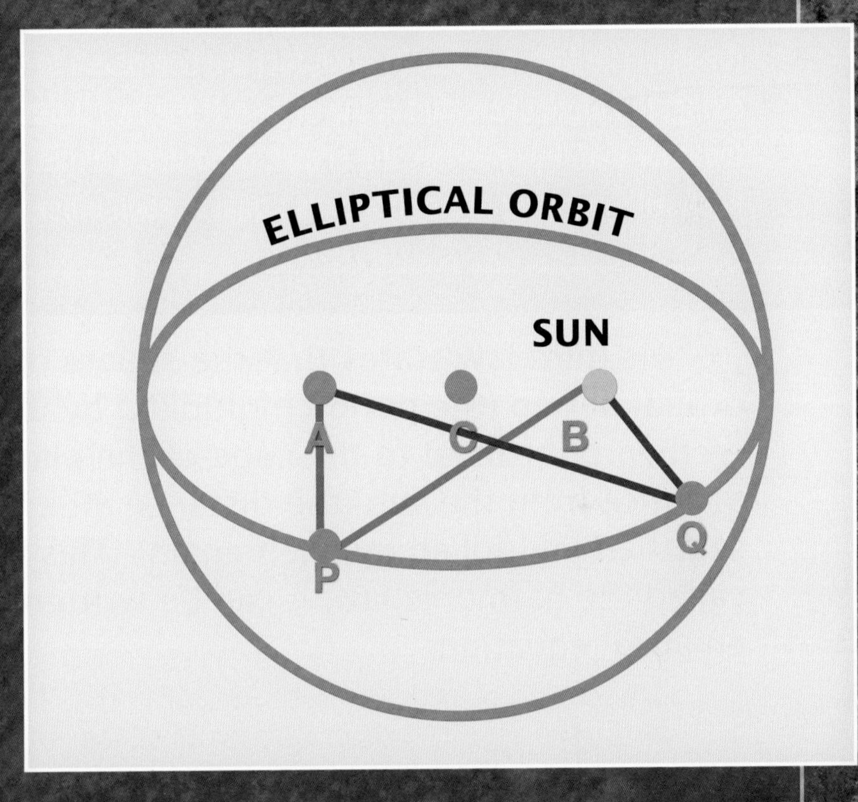

Not just any oval will do as an ellipse. If you choose any two points on the ellipse, such as the ones marked P and Q, the sum of the distances from P to A and P to B must be equal to the sum of the distances from Q to A and Q to B. It is easy to draw an ellipse on a piece of cardboard by sticking in two thumbtacks an inch or two apart near the center of the cardboard. Their positions will be the two foci of the ellipse. Then, make a loop of string around both thumbtacks. The loop should be a few inches longer than the distance between A and B when it is stretched to its limit. But it should be short enough that when you stretch it in any direction, it stays on the cardboard. Now stretch the loop with a pencil, and keeping it tight, draw an oval. That oval is an ellipse.

If the pencil point represents a planet, then one of the thumbtacks is the sun. The farther apart the two foci are compared to the length of the string, the more elongated is the ellipse. If the two thumbtacks are so close together that they overlap, the ellipse becomes a circle. A circle is an ellipse with its major and minor axes equal to each other.

His third law states that the square of a planet's orbital period (the period multiplied by itself) is directly proportional to the cube of the planet's average distance from the sun (the distance multiplied by itself then multiplied by itself again). This is sometimes called the harmonic law. It can be written as a mathematical equation:

$$P^2 = d^3$$

Here, P represents a planet's orbital period, the time in Earth years that it takes for the planet to go once around the sun. The d represents the planet's average distance from the sun measured in astronomical units or AU. One AU is the average distance between Earth and the sun.

OBSERVING WITH A TELESCOPE

In the same year that Kepler's *New Astronomy* was first published, Galileo Galilei began using a new tool, the telescope, in a way that its inventor had not thought of. He learned how to grind lenses and made his own telescopes, which were more and more powerful. And instead of using his telescope to see distant objects on Earth, he turned it toward the sky. By the fall of that year, Galileo had built a telescope that magnified objects twenty times.

He observed the moon in all of its phases over several weeks; he sketched what he saw. Though the church and many astronomers still believed that heavenly bodies have to be perfect, Galileo saw a surface filled with craters and mountains. Then in January 1610, Galileo observed four moons orbiting Jupiter and described his findings in *Sidereus Nuncius* (*The Sidereal Messenger*). Galileo later discovered the rings of Saturn and the fact that Venus has phases much like Earth's moon.

These discoveries were revolutionary. The surface of the moon showed that heavenly bodies are not necessarily perfect. The moons of Jupiter showed that there was more than one center of motion in the universe. The phases of Venus showed that the planet

TESTING KEPLER'S THIRD LAW

In the equation for Kepler's third law, the orbital period (*P*) is squared. The square of a number is that number multiplied times itself. In other words: *P* times *P*. In the equation, the distance (*d*) is cubed. This means *d* times *d* times *d*.

For Earth, P=1 year and d=1 AU, so the equation states the obvious: 1 squared is equal to 1 cubed. For Mars, d=1.52 AU and P=1.88 years, rounded off to the second decimal place. That gives $P^2=3.51$ and $d^3=3.53$. The small difference between those numbers comes from the rounding. When we use the best measurements of modern astronomy for every planet in the solar system, P^2 and d^3 are always equal to each other.

revolved around the sun in an orbit closer than Earth's, strengthening the heliocentric theory.

Because Galileo supported the idea of a heliocentric universe, leaders of the Catholic Church called him to Rome to defend himself. Cardinal Robert Bellarmine ordered him to keep quiet about his discoveries. Bellarmine instructed him "not to hold, teach, or defend" the Copernican theory "in any way whatever, either orally or in writing."

In 1624, Galileo returned to Rome and met with Pope Urban VIII. The pope gave Galileo permission to write a book, but he would allow it only on one condition: that Galileo treat the theories as if they were speculation, not scientific fact. By receiving the pope's permission, and agreeing to the terms, Galileo avoided the charge of heresy. In 1630 he completed *Dialogo sopra i due massimi sistemi del mondo, tolemaico e copernicano* (*Dialogue Concerning the Two Chief World Systems, Ptolemaic & Copernican*). Reading it today, there is no question that Galileo favored the Copernican system. But since it was written as a discussion between two fictitious characters, he had fulfilled the terms of his agreement.

Upon completion, Galileo sent the manuscript to a censor in Rome to check it for any signs that it went against the church's teachings. Because of an outbreak of the plague, communication was slow between Florence, where Galileo was, and Rome. So

GALILEO GALILEI WAS THE FIRST PERSON TO TURN A TELESCOPE TOWARD THE HEAVENS. HIS DISCOVERY OF THE PHASES OF VENUS SHOWED THAT IT ORBITED THE SUN CLOSER THAN EARTH, STRENGTHENING COPERNICUS'S HELIOCENTRIC THEORY. HE ALSO DISCOVERED FOUR MOONS ORBITING JUPITER, SHOWING THAT NOT ALL HEAVENLY MOTION IS EARTH-CENTERED. THESE DISCOVERIES PUT HIM AT ODDS WITH THE DOCTRINE OF THE POLITICALLY POWERFUL CATHOLIC CHURCH THAT EARTH WAS THE CENTER OF THE UNIVERSE. AS A RESULT, HE SPENT THE LAST TEN YEARS OF HIS LIFE UNDER HOUSE ARREST.

to speed up the process, Galileo requested that a censor in Florence read the manuscript. Though the censor had many criticisms of the book, he allowed it to be published, but only after Galileo wrote a clear introduction stating that the book was a completely fictional account. As such, the *Dialogue* first appeared in Florence in 1632.

Galileo died ten years later in 1642, having been subject to house arrest ever since the publication of the *Dialogue*. That same year, Newton was born, and he would carry forward the Scientific Revolution launched by Copernicus and continued by Kepler and Galileo.

CHAPTER THREE

NEWTON CONTINUES THE REVOLUTION

After Kepler and Galileo published their revolutionary books, few people doubted that Earth and the planets followed elliptical orbits around the sun. But what was there about the sun that held the planets in place? And why were the orbits elliptical? Those answers did not come easily. Solving those riddles of nature would take the rise of a rare genius, Isaac Newton.

Newton's brilliance was apparent to his professors and classmates at Cambridge from the beginning. Newton's friend William Whiston once said, "Sir Isaac, in mathematics, could sometimes see almost by intuition." Newton quickly mastered the works of the most advanced mathematicians of his day, but his curiosity for the discipline was not satisfied. He saw the world around him in ways

ISAAC NEWTON IS KNOWN FOR HIS BRILLIANCE IN MATHEMATICS. BUT HIS MATHEMATICS ALWAYS HAD ROOTS IN SCIENTIFIC OBSERVATIONS AND EXPERIMENTS. HERE, HE IS SHOWING CHILDREN THE PHENOMENON OF STATIC ELECTRICITY.

that no one else did, and he realized that not only the heavens but also many ordinary situations required new mathematical techniques to be understood fully.

Algebra was useful for figuring out the answers to simple equations. Geometry was good for dealing with shapes and spatial relations. But they were not sufficient to describe the constantly changing speed and direction of a moving object, such as a planet orbiting the sun or a tossed ball on Earth. A different form of mathematical analysis was needed, and Newton, at the age of only twenty-three, came up with it.

MOTION AND THE MATHEMATICS OF CHANGE

Newton called his idea "fluxions." It is now called calculus. Calculus deals with calculations of quantities or properties of an object that are constantly changing, whether it is the speed and position of a thrown ball or the motion of any object in the universe.

Imagine the curving path of a ball being shot from a cannon. The faster it leaves the barrel, the farther it goes before striking the ground. With a strong enough charge in the cannon, the ball could go fast enough so the curve of its path would match the curve of the Earth. It would still be falling toward the Earth, but it would never reach the ground. It would be in orbit, just like the moon. In fact, if it didn't fall, it would leave Earth forever. Newton realized that whatever made objects fall to the ground also kept the moon in its orbit around Earth and kept the planets in their orbits around the sun.

But how could he do calculations of such motion when it is changing from one instant to the next? The method of fluxions, which turned constant change into a series of infinitesimal changes, was the answer. By 1669 Newton had developed his method well enough to describe it in a paper called *De Analysi per Aequationes Numero Terminorum Infinitas* (*On Analysis of Equations with an Infinite Number of Terms*). The paper was circulated among a select group of scholars and immediately sparked interest. Its success made Newton famous as a leading mind in mathematics.

LIGHT, COLOR, AND OPTICAL INSTRUMENTS

Around this same time, Newton was also conducting experiments in an area of study that seemed unrelated—light. It began with the prism experiments described previously, but it continued for many years.

Before Newton's experiments, many people believed that colors were modifications of white light. Newton's work with prisms showed instead that white light is really a combination of all the colors in the

NEWTON'S EXPERIMENTS WITH A GLASS PRISM LED HIM TO REALIZE THAT WHITE LIGHT CONTAINS ALL COLORS. EACH COLOR REFRACTS (BENDS) DIFFERENTLY WHEN IT CROSSES A BOUNDARY FROM AIR TO GLASS AND BACK AGAIN, SO THE TRIANGULAR SHAPE OF THE PRISM CREATES A SPECTRUM. NEWTON DISAGREED WITH HUYGENS AND ROBERT HOOKE ABOUT WHAT CAUSED REFRACTION. NEWTON SAID IT WAS BECAUSE LIGHT IS A STREAM OF PARTICLES, BUT HOOKE AND HUYGENS SAID IT WAS BECAUSE LIGHT IS A WAVE.

spectrum. In the same way that shining red and green lights on a screen produce yellow, shining all the colors in the spectrum makes white.

A prism breaks white light up into its colors because of the phenomenon of refraction. When light strikes a surface between two transparent substances, like air and glass, at an angle, it changes its direction. Different colors change direction by different amounts, so in Newton's experiment, the colors spread out inside the glass of the prism. The colors then reached the other surface of the prism at a different angle, so they didn't blend back together.

Refraction is also what makes a lens work, so Newton realized that telescope lenses would focus different colors at different places. That would make an image blurry. As a result, he decided to try to build a telescope with a curved mirror to form the image instead of a lens. As noted previously, reflecting telescopes also had other advantages.

THE NATURE OF LIGHT

None of Newton's experiments with light could ever answer the question of what light was made of. Hooke and Huygens compared refraction with the way water waves change direction when they strike an obstacle. Newton argued that light was made of tiny particles called corpuscles. He imagined refraction to be the change in direction that happens when a ball rolls down a wide ramp and reaches level ground. If it is not going straight down the middle but rather at an angle (say from the upper right to the lower left), it changes direction at the bottom.

There was no way to settle the argument at the time. If light was waves, the crests had to be so close together that no one saw them. If light was a stream of corpuscles, they had to be too tiny to be seen. So the rivals kept arguing about the nature of light as long as they lived.

THE MATHEMATICS OF UNIVERSAL GRAVITATION

Newton had realized quite early that the gravitational attraction that caused objects to fall to Earth also kept the moon and the planets in their orbits. He considered it a universal law of nature, but he did not immediately figure out what mathematical formula to use to describe it. When he finally did, it was due in part to correspondence he had with Robert Hooke.

The law states that any two objects in the universe attract each other gravitationally. The force depends on the mass, or the amount of matter, of each body. The more mass each body has, the stronger the force of gravity between them is. The relationship between mass and gravitational force follows what mathematicians call a direct proportion. Double the mass of one body means double the gravitational force between it and any other body. Triple the mass means triple the force, and so on.

The gravitational force also depends on how far apart the objects are. It is stronger when they are close and weaker when they are farther apart. Specifically, it follows what is known as an inverse-square relationship, which Hooke suggested in letters to Newton in 1679. Double the distance means the attraction is one-fourth as strong (one-half times one-half). Triple the distance means the attraction is only one-ninth as strong (one-third times one-third), and so forth.

By early 1684, English astronomer Edmond Halley (1656–1742) was having similar conversations about gravity with both Hooke and architect Sir Christopher Wren (1632–1723). To Halley's excitement, Hooke said that he had come to the same conclusion that the strength of gravity decreased as the square of the distance. The problem, of course, was showing that the formula predicted the motion of the planets, which no one had ever done.

As an incentive, Wren offered to give a valuable book to whichever of the two men could solve the problem. Hooke, in his proud manner, said that he

BRITISH ASTRONOMER EDMOND HALLEY RECOGNIZED THAT THE MATHEMATICAL FORM OF NEWTON'S LAW OF UNIVERSAL GRAVITATION COULD EXPLAIN WHY KEPLER'S LAWS WERE TRUE. HE ENCOURAGED NEWTON TO PUBLISH HIS FORMULA, BUT NEWTON NEEDED TO DO MORE CALCULATIONS BEFORE HE WAS SATISFIED. EIGHT YEARS LATER, NEWTON FINALLY PUBLISHED *PHILOSOPHIAE NATURALIS PRINCIPIA MATHEMATICA* (*MATHEMATICAL PRINCIPLES OF NATURAL PHILOSOPHY*), OFTEN CONSIDERED THE GREATEST MASTERPIECE OF SCIENCE EVER WRITTEN.

already knew the answer but was keeping it a secret until he decided to make it public. In turn, Halley called upon Newton to see if he could solve the problem. Halley boarded a train from London to Cambridge in August of that year. Having met Newton only once before, Halley didn't know what to expect because of the scientist's reputation as a recluse. But Newton was happy to see him.

After talking for a while, Halley asked Newton the question he had come a long way to ask: What kind of curve "would be described by the planets supposing the force of attraction towards the sun to be reciprocal to the square of their distances from it?" To Halley's surprise, Newton had the same answer as he, Hooke, and Wren had concluded—an ellipse. Halley was greatly surprised when he asked Newton how he had come to that conclusion. Newton replied simply, "I have calculated it."

Halley was eager to see the answer and for Newton to share his calculation with the world. He would eventually get his wish—and much more—but it would take eight years before Newton published his calculation in the *Principia*.

CHAPTER FOUR

THE PATH TO THE *PRINCIPIA*

Halley knew that Newton had come up with a breakthrough, so he was disappointed when Newton told him he had lost his notes. Halley made Newton promise to rewrite them and send them to him once they were complete. Eleven months later, in November 1684, Newton delivered to Halley the first copy of *De Motu Corporum in Gyrum* (*On the Motion of Revolving Bodies*).

When Halley read *De Motu*, he was thrilled. This was the first paper published on the relationship between planetary motion and the force of gravity that guides that motion. Halley so believed in its importance that he visited Newton again to ask if he would present the paper before the Royal Society as well as publish it. Newton agreed.

De Motu was well received by the Royal Society, which urged Halley to persuade Newton to publish as soon as possible. For the next year and a half, Newton prepared *De Motu* for publication. In April 1686, Newton presented the first third of the manuscript to the Royal Society. It was so clearly important that the Royal Society offered to pay to publish it, but soon the members realized that the society was out of money. In the end, Halley financed the project.

In its final form, *De Motu* was published with the title *Philosophiae Naturalis Principia Mathematica* (*Mathematical Principals of Natural Philosophy*), or simply the *Principia*. Today, many scientists believe that the *Principia* is the most important text ever written in the history of science. The *Principia* describes the physics of motion through three basic laws, and in the eyes of many people, it brought mathematical order to the universe.

A MASTERPIECE IN THREE BOOKS

The *Principia* was published in July 1687 and comprised three books. Book I explains the concept of motion without friction. Book II concerns the motion of fluids and the effect of friction on solid bodies moving in fluids. Book III is considered profound and what made the *Principia* an important text. The third book covers Newton's famed three laws of motion and the law of universal gravitation.

Newton's first law states, "Every body [object] continues in its state of rest, or of uniform motion in a straight line, unless it is compelled to change that state by forces impressed upon it." This was not an original idea; Galileo first thought of it nearly fifty years earlier. But Galileo had applied it only to objects moving on Earth. Newton extended it to the behavior of all matter, including the movement of the planets.

According to the first law, the planets' orbital motion required a force to be acting on them. Otherwise they would move in a continuous straight line out into space. The second law explains how that force changes a planet's speed and direction. It states, "The change of motion is proportional to the motive force impressed; and is made in the direction of the straight line in which that force is

[488]

tertiam, X longitudinem quam Cometa toto illo tempore ea cum velocitate quam habet in mediocri Telluris à Sole diſtantia, deſcribere poſſet, & tV perpendiculum in chordam $T\tau$. In longitudine media tB ſumatur utcunque punctum B, & inde verſus Solem S

ducatur linea BE, quæ ſit ad Sagittam tV ut contentum ſub SB & St quadrato ad cubum hypotenuſæ trianguli rectanguli, cujus latera ſunt SB & tangens latitudinis Cometæ in obſervatione ſecunda ad radium tB. Et per punctum E agatur recta AEC, cujus partes AE, EC ad rectas TA & τC terminatæ, ſint ad invicem ut tempora V & W: Tum per puncta A, B, C, duc circumferentiam circuli, eamque biſeca in i, ut & chordam AC in I. Age occultam Si ſecantem AC in λ, & comple parallelogrammum $iI\lambda\mu$. Cape $I\sigma$ æqualem $3I\lambda$, & per Solem S age occultam $\sigma\xi$ æqualem $3S\sigma + 3i\lambda$. Et deletis jam literis A, E, C, I, à puncto B verſus punctum ξ duc occultam

THIS DIAGRAM FROM NEWTON'S *PRINCIPIA* DESCRIBES HOW TO CALCULATE THE ORBIT OF A COMET FROM THREE MEASUREMENTS OF ITS POSITION IN THE SKY AT DIFFERENT TIMES. TODAY, ASTRONOMERS USE A SIMILAR TECHNIQUE TO DETERMINE THE ORBIT OF NEWLY DISCOVERED OBJECTS, SUCH AS COMETS, ASTEROIDS, AND OBJECTS IN THE KUIPER BELT—THE DISTANT REGION OF THE SOLAR SYSTEM THAT CONTAINS THE FAMOUS DWARF PLANET PLUTO.

impressed." Simply stated, the reason the planets do not continue in a straight line is because they are continually pulled toward the sun, specifically by the sun's immense gravity.

Newton's third law states, "To every action there is always opposed an equal reaction: or, the mutual

FEELING HOW GRAVITY PRODUCES A PLANET'S ORBIT

For a planet orbiting the sun, gravity pulls in the direction from the planet toward the sun, but the planet's velocity is in a different direction, namely along the orbital path. Velocity is different from speed because it has a direction. Velocity can be represented by an arrow of a certain length in the direction of motion.

You can feel how that difference in direction between force and velocity produces a circular orbit by swinging a ball on a string. Notice that you are always pulling the ball inward with a steady force, while the direction of the ball's velocity keeps changing along its path. Your force produces a change in velocity every instant by adding a very small velocity arrow, pointing in the direction of your force, to the tip of the first one. The new velocity is an arrow from the tail of the large arrow to the tip of the small one. Its direction is different, but its length can be larger, smaller, or the same depending on the direction of the force.

To feel the same thing for planets in elliptical orbits, you would have to allow the string to change length and adjust the force steadily to match the inverse-square law of gravity. That would be very difficult to do because, unlike gravity, the force of tension in the string does not naturally have an inverse-square relationship to the string's length.

action of two bodies upon each other are always equal, and directed to contrary parts." In more familiar language, the third law states that forces always act in equal and opposite pairs between two bodies. Whenever one body pushes or pulls on another, the other body is pushing or pulling with the same amount of force on the first in the opposite direction along the line between them.

For example, Earth pulls on the moon, and in turn, the moon pulls on Earth with an equal amount of force. The reason we say the moon orbits Earth rather than the other way around is due to the fact that Earth is so much more massive than the moon. Therefore, Earth moves

THE EARTH-MOON SYSTEM ILLUSTRATES NEWTON'S THREE LAWS OF MOTION AND HIS LAW OF UNIVERSAL GRAVITATION. BECAUSE OF INERTIA (THE FIRST LAW OF MOTION), THE MOON WOULD TRAVEL IN A STRAIGHT LINE AT CONSTANT SPEED, BUT GRAVITY PROVIDES A FORCE THAT CHANGES ITS MOTION INTO AN ORBIT. THAT ORBIT CAN BE CALCULATED USING THE SECOND LAW OF MOTION AND THE LAW OF GRAVITY. THE THIRD LAW STATES THAT THE MOON AND EARTH ATTRACT EACH OTHER BY GRAVITY AND ORBIT EACH OTHER. THAT CAUSES EARTH TO WOBBLE AROUND A POINT ABOUT THREE-QUARTERS OF THE WAY OUTWARD FROM ITS CENTER. THAT WOBBLE IS WHAT CAUSES EARTH'S TIDES.

less than the moon. But if you were an astronaut on Mars watching the motion of Earth and the moon, you would notice that they are rotating together around a point called their center of mass. That point is always on the line between their centers about three-quarters of the way from Earth's center to its surface. Earth would look like it is wobbling, and that wobbling is what causes the oceans' tides.

Newton's law of universal gravitation (Proposition VII of Book III of the *Principia*) states that every object in the universe is attracting every other object with a force pair that obeys the third law of motion. It states, "Every particle of matter attracts every other particle with a force proportional to the products of the masses and inversely proportional to the square of the distances between them."

That proposition declares that the gravity of every thing in the universe effects the movement and position of every other thing. Anything with mass, from the largest star to a grain of sand (even to the smallest subatomic particles, though Newton did not know that there were such things), interacts with any other body or particle with mass in the universe. Accepting that proposition demanded a change in thinking for many people. The heavens could no longer be viewed as different from Earth with different laws. The laws of motion and gravity were the same throughout the universe.

CHAPTER FIVE

UNDERSTANDING NEWTON'S LAWS

To many people, understanding Newton's laws of motion seems to be a challenge. But in fact, they are straightforward to explain by considering everyday phenomena. It is a matter of viewing those phenomena from the right perspective. Let's look at them one at a time.

NEWTON'S FIRST LAW: THE LAW OF INERTIA

In the *Principia*, Newton stated the first law of motion this way: "Every object persists in its state of rest or uniform motion in a straight line unless it is compelled to change that state by forces pressed upon it." This means that an object will continue to move at the same velocity (the same speed and in the same direction) unless an outside force causes that motion to change. If the speed is zero, the body is at rest, and it takes a force to set it in motion. That is the Law of Inertia, or resistance to changes in motion.

At first glance, that seems strange. Balls start to roll downhill and speed up as they go, but they slow to a stop on level ground. Cars slow down unless you give them gas. Objects dropped from a window speed up until they hit the ground. Baseballs follow arcs instead of straight lines. But if you think of those familiar phenomena in terms of Newton's Law of Inertia, you realize that forces are involved.

Rolling balls and cars slow down, even on level ground, because a familiar force

EVEN IF THEY DO NOT THINK ABOUT SCIENCE AS THEY SHOOT THE PUCK, HOCKEY PLAYERS ILLUSTRATE NEWTON'S FIRST AND SECOND LAWS OF MOTION. NEWTON'S FIRST LAW, THE LAW OF INERTIA, SAYS THAT AN OBJECT WILL CONTINUE TO MOVE AT THE SAME SPEED IN THE SAME DIRECTION UNLESS A FORCE ACTS ON IT. (IF THAT SPEED IS ZERO, IT IS AT REST AND WILL STAY AT REST.) BECAUSE ICE HAS VERY LITTLE FRICTION, A MOVING HOCKEY PUCK WILL SLIDE AT CONSTANT SPEED AND DIRECTION UNTIL A PLAYER HITS IT. THAT'S THE FIRST LAW.

WHEN A PLAYER HITS THE PUCK, THE BLADE OF THE HOCKEY STICK EXERTS A FORCE ON THE PUCK AND ACCELERATES IT, CHANGING ITS SPEED AND DIRECTION. THE FORCE AND ACCELERATION ARE MATHEMATICALLY RELATED TO EACH OTHER BY THE SECOND LAW. AFTER THE HIT IS OVER, THE PUCK IS SLIDING STEADILY IN A DIFFERENT DIRECTION AND SPEED, AGAIN OBEYING NEWTON'S FIRST LAW.

called friction acts against their movement. Friction results when objects move across each other. The objects can be solid, but they don't have to be. When a car moves through the air, it has to push the air aside, and that means the air is pushing back on it. To keep a car moving at constant speed, the engine has to create a force on the wheels to match that friction. To speed the car up, the engine has to create an even larger force. Balls rolling downhill, falling objects, and thrown baseballs change their velocities because gravity is acting on them.

So when you see an object changing its velocity (its speed, direction, or both), Newton's first law tells you that a force must be acting on it. And it usually isn't too hard to tell what that force is coming from. That not only applies to objects on Earth but also in space. In space, there is no air to slow a moving object down, so the moon would quickly fly away from Earth if it weren't for the force of gravity. The same is true of the planets and the sun. The moon orbits Earth and the planets orbit the sun because gravity keeps them from continuing to move in a straight line.

Even in deep space, objects move under the influence of gravity. For instance, the *Pioneer 11* spacecraft was launched in 1973 to study the asteroid belt, Jupiter, and Saturn. Its speed and direction have kept changing under the influence of the sun's powerful gravity ever since. It has also been influenced by the gravity of all the planets, but that influence has been much smaller than the sun's except when it was came fairly close to Jupiter and Saturn. It is now reaching the edge of the solar system, where the stream of particles called the solar wind stops.

It will now continue to coast away from the solar system in a region of space between the stars that has so little matter that nothing will change its speed except an ever-weakening gravity from the sun. In about four million years, it will pass close enough to another star, in the constellation of Aquila, to have its path influenced by that star's gravity much more than the sun's.

NEWTON'S SECOND LAW: THE LAW OF FORCE AND ACCELERATION

When you think about inertia, you realize that it is much easier to change the velocity of a baseball than a car. That is because the car has more mass. We commonly speak of weight rather than mass. But scientists prefer to speak of mass because weight depends on the strength of gravity. A liter of water on Earth weighs about 2.2 pounds. On the moon, where gravity is weaker, it weighs only one-sixth as much. But its mass is one kilogram in both places. Mass is a measure of a body's inertia.

Newton's second law of motion is an equation that describes the way a body's velocity changes under the influence of a force. Simply stated, it is this: The change in motion of a body is proportional to the force pressed upon it, or in mathematical terms, it looks like this:

$$F = ma$$

Here, F is the force acting on a body that has a mass m. The force causes the body's velocity to change at a rate we call the acceleration, or a in the equation. Like velocity, acceleration has a direction as well as a size, or magnitude. The equation tells us that the direction of the acceleration is the same as the force, and that to calculate the force you multiply the body's mass by its acceleration.

For instance, a body on Earth that has only the force of gravity acting on it changes its vertical motion in the downward direction at a steady rate. If we drop a stone out of a high window, it starts out with a downward speed of zero. After one second, it will be traveling downward at about 32 feet or 9.8 meters per second (not counting air friction). After two seconds, its speed will be 64 feet or 19.6 meters per second. After three seconds, its speed will be 96 feet or 29.4 meters per second, and so forth. Because the force acting on it is steady, its acceleration is steady. Its velocity is changing at a rate of 32 feet or 9.8 meters per second every

THE MISUNDERSTOOD THIRD LAW

Newton's third law is often misunderstood because people forget that the "action" and "reaction" must act between the same pair of bodies. Consider a 100-pound (45 km) person standing on the sidewalk. Earth's gravity pulls that person downward with a 100-pound force, but the person doesn't accelerate because the sidewalk is pushing upward on the person's shoes with a 100-pound contact force. The two forces on the person, gravity and the contact force, are equal and opposite. But they can't be an action-reaction pair since they both act on same body, namely the person. Because they are equal to each other and in opposite directions, the net force on the person is zero. Thus according to the second law, not the third, the person does not accelerate.

NEWTON'S THIRD LAW STATES THAT FORCES ALWAYS ACT IN ACTION-REACTION PAIRS BETWEEN TWO BODIES. IN THE EXAMPLE SHOWN HERE, THE ACTION AND REACTION FORCES ACT BETWEEN THE TOE OF THE SKATER'S SHOE AND THE SKATEBOARD. THE FORCE LABELED ACTION IS THE FORWARD PUSH OF THE SHOE ON THE SKATEBOARD, WHICH CAUSES THE SKATEBOARD TO SPEED UP. THE FORCE LABELED REACTION IS THE SKATEBOARD'S BACKWARD PUSH ON THE SHOE. BEFORE PUSHING, THE SKATER IS LEANING FORWARD, BUT THE BACKWARD PUSH ON HIS SHOE CAUSES HIM TO MOVE SLIGHTLY SLOWER THAN THE SKATEBOARD AND THUS TO STAND UP STRAIGHTER AS THE SKATEBOARD CATCHES UP TO HIS BODY.

second in the downward direction. That means its acceleration is 32 feet or 9.8 meters per second per second.

In the case of a falling body, *F* in the equation is the force of gravity, or the body's weight. In everyday units, this is measured in pounds, but in scientific units or the metric system, it is measured in units called newtons, in honor of the creator of the equation. If a body has a mass of one kilogram, the equation for Newton's second law tells us that it weighs 9.8 newtons on Earth. If it has a mass of 2 kilograms, its weight is 19.6 newtons (4.4 pounds).

Newton's second law applies to all situations, not just falling bodies. Suppose you are riding a bike on level ground and the mass of you and your bike together is 60 kilograms (corresponding to a weight of about 132 pounds). You pedal as hard as you can for 6 seconds and reach a speed of 9 meters per second (20 miles per hour). That means your acceleration is 9/6 = 1.5 meters per second per second, and the forward force on the bike of your pedaling is 60 times 1.5 or 90 newtons (20 pounds). Actually, you would be pedaling with an even greater force because you have to overcome friction in the bicycle and air resistance that are forces acting against your efforts.

THE THIRD LAW

Newton's third law states that forces always act in equal and opposite pairs between two bodies.

Whenever one body pushes or pulls on another, the other body is pushing or pulling with the same amount of force on the first in the opposite direction along the line between them. The easiest example to illustrate Newton's third law is a rocket engine. In the rocket engine, the fuel burns inside a chamber with an exhaust vent at the back end. The burning creates a very high pressure inside the chamber, creating a very large force that pushes the exhaust gas through the vent at high speed. By Newton's third law, that means the exhaust gases are pushing forward on the chamber at the same time. That forward push accelerates the rocket.

You can create a similar effect very easily with a balloon. Blow it up as much as you can and pinch the balloon's neck tightly without tying it. Then release the neck. The high pressure in the balloon forces the air backward through the neck, and the backward rushing air pushes the balloon like a rocket.

One unexpected example of the third law in action is the way a bicycle or car moves forward. The weight of the car or bicycle pushes the tires firmly against the pavement. The engine or the bicyclist's pedaling starts to rotate the tires, which push backward against the ground. The third law says that the ground must be pushing forward on the tires at the same time. The energy to turn the wheels comes from the bicyclist or the engine, but it is the ground that gives the bike or car its forward push!

CHAPTER SIX

ON NEWTON'S SHOULDERS

When Newton spoke of standing on the shoulders of giants, he knew that he was part of a long tradition of knowledge and learning. The people who would follow and stand on his shoulders were certain to see things that were out of the reach of scientists of his time. What makes Newton's work so remarkable is that much of it is still the foundation of science nearly four centuries after his death. Modern mechanics is more a refinement than a correction to his work. Many of the questions that he asked in the seventeenth century are still being pursued in the twenty-first.

THE RETURN OF THE COMET

The ideas in Newton's *Principia* led directly to the discovery of Halley's comet. Comets had always

THE MOST FAMOUS OF ALL COMETS IS NAMED AFTER EDMOND HALLEY. HALLEY DID NOT DISCOVER THE COMET, BUT HE WAS THE FIRST TO COMPUTE ITS ORBIT USING TECHNIQUES DESCRIBED IN NEWTON'S *PRINCIPIA*. IT WAS THEN CALLED THE GREAT COMET OF 1682. HIS CALCULATIONS SHOWED THAT IT WAS IN A LONG ELLIPTICAL ORBIT WITH A PERIOD OF BETWEEN 75 AND 76 YEARS. HE REALIZED THAT THIS COMET HAD ALSO BEEN SPOTTED IN 1531 AND 1607 AS WELL AS OTHER EARLIER DATES IN HISTORY. WHEN IT RETURNED AS PREDICTED IN 1758, IT BECAME KNOWN AS HALLEY'S COMET.

been somewhat of a mystery to scientists because of their irregular movement across the sky. Halley realized that like planets, comets move in elliptical orbits. But unlike planets, their orbits are very elongated.

Halley determined the elliptical orbit of a great comet in 1682, and he realized that it would have a period of between seventy-five and seventy-six years. Halley concluded that this comet had been spotted at least twice before, once in 1531 and again in 1607. Then, looking at historical records, he realized that it had been seen many times before. Its most famous appearance occurred in 1066, right before the Battle of Hastings. Historians through the ages have recorded that William of Normandy (also known as William the Conqueror) claimed the comet was a sign that he would be victorious in battle over King Harold.

Based on that information, Halley predicted that the comet would return around the year 1758. It did return in 1758, on Christmas Day, 116 years after the birth of Isaac Newton. And it has been known ever since as Halley's comet.

A GRAVITATIONAL GUIDE TO THE PLANETS

Immediately following the publication of the *Principia*, Newton and others put his newfound discoveries to work. For several years he had been trying to match Saturn's orbit to an ellipse, but he could never quite get observations to agree with his calculations. Newton realized that Jupiter was large enough and close enough to Saturn that its gravity pulled Saturn out of the perfect ellipse that would be caused if the sun was the only influence on the ringed planet. Newton called Jupiter's influence on Saturn's orbit "the three-bodied problem."

This was far from the last time that the gravity of one or more planets was found to influence the orbit of others. On March 13, 1781, William Herschel

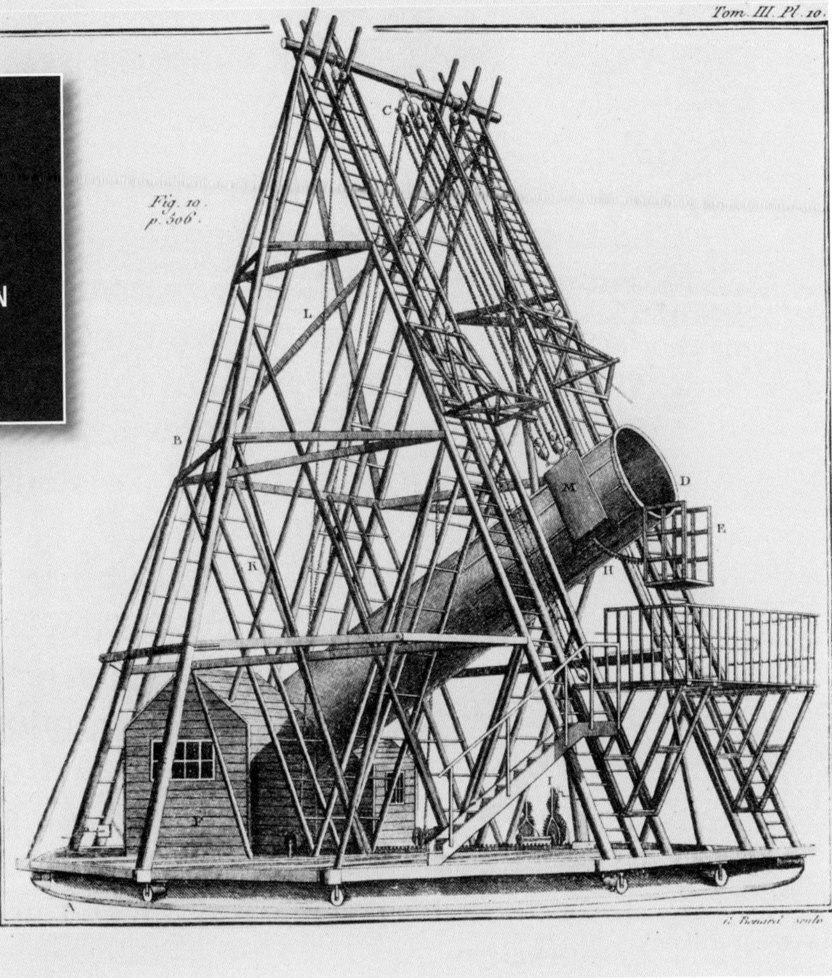

AFTER ASTRONOMER WILLIAM HERSCHEL DISCOVERED URANUS IN 1781, HE WAS APPOINTED AS PRIVATE ASTRONOMER TO KING GEORGE III OF ENGLAND. HE THEN BUILT THE LARGEST TELESCOPE OF ITS TIME, SHOWN HERE.

(1738–1822) discovered Uranus. After several years of measurements, astronomers realized that its orbit did not match their predictions even considering the influence of Jupiter and Saturn. They concluded that there must be a planet farther from the sun that no one had seen. Scientists John Couch Adams (1819–1892) and Urbain Jean Joseph LeVerrier (1811–1877) applied Newton's laws, made the calculations, and predicted where it could be found. On September 23, 1846, on their first night of observation, scientists at the Berlin Observatory discovered the planet Neptune.

A similar process led to the discovery of Pluto in 1930. But additional observations by telescopes led astronomers to realize that Pluto was much smaller than first thought. Also new measurements of the outer planets' masses by two Voyager spacecraft in the 1980s showed that the orbits of Uranus and Neptune could be calculated without the need for a large planet at Pluto's distance. We now know that Pluto is one of the largest dwarf planets in a distant region of the solar system called the Kuiper Belt.

Pluto, at least, proved to be an actual body, but another predicted planet was never found for a very different reason. In 1859, LeVerrier noticed that the orbit of Mercury seemed to be influenced by another planet even closer to the sun. Astronomers looked for it until at least 1915, when Albert Einstein published a paper describing his general theory of relativity. In it, he transformed Newton's law of universal gravitation.

According to Einstein's theory a body warps measurements of space and time in its vicinity, causing what can be observed as a gravitational attraction that follows Newton's formula almost, but not quite, exactly. The difference between Einstein's theory and Newton's is large enough to detect in measurements near a very large body—for example in an orbit like Mercury's that is close to the sun. General relativity, not a missing planet, explains the difference of Mercury's orbit from what Newton's laws alone would predict.

SEEING THE LIGHT ABOUT LIGHT

As noted previously, one of the great disputes between Newton and Hooke regarded the nature of light. Is it a wave or is it a stream of particles? In 1801, Thomas Young (1773–1829) performed an experiment that seemed to answer the question once and for all. He passed a light beam through two closely spaced slits and let it fall on a screen behind them. If light were a stream of particles, he would have expected to see two spots of light. Instead he saw a pattern of repeating light and dark bands, similar to the pattern of crests and valleys that forms when two waves meet.

Young's experiment persuaded scientists that light was a wave phenomenon. Then in 1861–1862, James Clerk Maxwell (1831–1879) published a series of scientific papers that described the electric and magnetic properties of matter and space in four equations. Those equations predicted the existence of electromagnetic waves that traveled through space at a certain

YOUNG'S EXPERIMENT IN A POND

You can try a simple version of Young's experiment for yourself. Go to a still pond or an inflatable swimming pool and gently drop a pebble into the water. It creates a series of spreading waves. Now drop two pebbles at the same time a few feet apart. Watch what happen when the waves meet. You will see a pattern of lines where two wave crests or two wave valleys cross each other.

speed that could be calculated from two constants, one in the equation describing the force between electric charges and the other in the equation describing the force between magnetic fields. Remarkably, that speed matched the speed of light.

That strengthened the idea that light was a type of wave—specifically an electromagnetic wave with a very small wavelength. Red had the longest wavelength of all colors, and violet had the shortest.

Newton's idea of light particles seemed to be dead, that is until the year 1900. That was when Max Planck (1858–1947) tried to analyze the spectrum of light coming from a heated body. Using the idea that electromagnetic waves can have any height, he calculated a spectrum that matched measurements very well in the invisible but detectable

IN 1861–62, PHYSICIST JAMES CLERK MAXWELL DESCRIBED THE RELATIONSHIPS BETWEEN ELECTRICITY AND MAGNETISM IN A SET OF FOUR EQUATIONS. THOSE EQUATIONS PREDICTED ELECTROMAGNETIC WAVES THAT TRAVEL AT THE SPEED OF LIGHT. THAT PREDICTION, WHEN COMBINED WITH A FAMOUS EXPERIMENT BY THOMAS YOUNG IN 1801, SEEMED TO SETTLE THE ARGUMENT OF WHETHER LIGHT WAS A WAVE OR A STREAM OF PARTICLES IN FAVOR OF HUYGENS AND HOOKE. BUT IT TURNED OUT NOT TO BE THE FINAL WORD.

colors with wavelengths longer than red called infrared. But his calculations were a complete mismatch in the ultraviolet region (wavelengths shorter than violet).

Planck made a guess that electromagnetic waves like sand than of a liquid. They had a minimum value (a quantum), with smaller grains for red light than for blue. When he put that idea into his calculations, his spectrum matched perfectly. Could light be a stream of particles after all? In 1905, Einstein used Planck's idea to explain another odd phenomenon called the photoelectric effect.

Now there were two famous results that said that light sometimes appeared to act like a wave (Young's experiment), and sometimes it appeared to be a stream of particles (the photoelectric effect). Both were correct! That meant that scientists were asking the wrong question about whether light is a wave or a particle phenomenon. The answer is not either-or but both.

In the two decades that followed, other scientists realized that things like electrons that they had also been calling particles might act like waves under some circumstances. That turned out to be true, and the branch of physics known as quantum mechanics was born. Most of the advances in modern technology, like lasers, electron microscopes, and computers, would not have been invented without an understanding of quantum mechanics.

And all of those discoveries and inventions would not have come to pass if we had not had the shoulders of Newton to stand on.

TIMELINE

1543 Polish astronomer Nicolaus Copernicus publishes *De Revolutionibus Orbium Coelestium* (*On the Revolutions of Heavenly Spheres*), which argues that the sun and not Earth is the center of the universe.

1609 Johannes Kepler publishes his book *Astronomia Nova* (*New Astronomy*), which states that the planets move in elliptical rather than circular orbits around the sun.

1610 Galileo observes four moons orbiting around Jupiter in January and describes his findings in his book *Sidereus Nuncius* (*The Sidereal Messenger*).

1630 Galileo completes *Dialogo sopra i due massimi sistemi del mondo, tolemaico e copernicano* (*Dialogue Concerning the Two Chief World Systems, Ptolemaic & Copernican*).

1642 Isaac Newton is born in Woolsthorpe, Lincolnshire, England, on Christmas Day.

1661 Newton arrives at Trinity College.

1664 Newton publishes his book of philosophical questions titled *Quaestiones Quaedam Philosophicae* (*Certain Philosophical Questions*)

1665 Newton leaves Trinity College.

1669 Newton summarizes his findings on calculus in the paper titled *De Analysi per Aequationes Numero Terminorum Infinitas* (*On Analysis of Equations with an Infinite Number of Terms*).

1675 Newton writes *An Hypothesis Explaining the Properties of Light.*

1678 Newton suffers from a nervous breakdown.

1679 Newton completes his outline on the three laws of motion and universal gravitation, ideas that would lead to his landmark work *Philosophiae Naturalis Principia Mathematica* (*Mathematical Principals of Natural Philosophy*), commonly known as the *Principia.*

1682 Edmond Halley uses Newton's laws of gravity and motion to calculate the orbit of a great comet and predicts its return seventy-six years later.

1684 Newton delivers to Halley the first copy of *De Motu Corporum in Gyrum* (*On the Motion of Revolving Bodies*).

1686 Newton presents to the Royal Society the first third of his manuscript *De Motu Corporum in Gyrum* (*On the Motion of Revolving Bodies*).

1687 The *Principia* is published.

1696 Newton accepts the post of warden of the British mint.

1703 Newton is elected president of the Royal Society.

1705 Newton is knighted for his scientific achievements by Queen Anne of Great Britain.

1706 Newton publishes a Latin edition of *Opticks.*

1718 Newton publishes an English edition of *Opticks.*

1727 Newton dies on March 31, 1727, in London.

1758 Halley's comet appears as predicted.

1781 Uranus is discovered.

1801 Young's experiment demonstrates the wave properties of light.

1846 Neptune predicted using Newton's law and discovered where it was expected.

1861–1862 Maxwell's electromagnetic equations predict electromagnetic waves that travel at the speed of light.

1900 Planck invents the quantum as a mathematical idea to explain the spectrum of a heated body.

1905 Einstein proposes that the light quantum actually exists as a particle and uses it to explain the photoelectric effect.

1905–1915 Einstein develops the special and general theories of relativity that lead to a modification of both Newton's laws of motion and Newton's law of universal gravitation.

1910S–1930S Physicists develop quantum mechanics, which adapts Newton's laws to the new understanding that wave properties and particle properties cannot be completely separated.

1950S–2000S Understanding of quantum mechanics leads to the electronics revolution that transforms communication and computation.

GLOSSARY

ACCELERATION The rate of change of velocity.

CELESTIAL Of or from the sky.

COUNTERFEITER Someone who creates fake money in hopes of profiting from its use.

DEFERENT A circle centered on the point halfway between Earth and the equant, used in Ptolemy's geocentric description of the universe as a planet's main motion.

EPICYCLE A circle around a point on the deferent, used to explain the paths of the sun, moon, and planets in Ptolemy's geocentric description of the universe.

EQUANT A point used in Ptolemy's geocentric description of the universe to explain the sun's changing speed around Earth at different times of the year. He stated that the sun changed the angle it made with the line between the equant and Earth at a constant rate.

FLUXIONS The name Newton gave to a branch of mathematics that he invented to deal with continuous changes. It eventually became known as calculus.

FORCE A push or pull exerted by one body on another.

FRICTION A force that resists movement of one body past another.

GEOCENTRIC A planetary system with Earth as its center.

INERTIA The tendency of matter to maintain its velocity (or state of rest).

LAW A scientific principle that applies broadly in a particular area of study.

MASS A measure of the inertia of an object.

MECHANICS The area of physics that deals with forces and motions.

OPTICS The area of physics that deals with the properties of light and its interaction with matter.

ORBIT The path one body takes around another.

SCIENTIFIC REVOLUTION A historical period during which major changes in scientific knowledge took place, usually defined to begin with Copernicus's proposed heliocentric model of the universe.

SPECTRUM A range of color formed when a beam of light is broken into its individual colors.

VELOCITY The rate of motion in a particular direction.

FOR MORE INFORMATION

American Astronomical Society (AAS)

2000 Florida Avenue NW, Suite 400

Washington, DC 20009

(202) 328-2010

Web site: http://www.aas.org

The mission of the AAS is to enhance and share humanity's scientific understanding of the universe. It sponsors meetings and publishes journals for scientists and also publishes information for the public, educators, and people interested in careers in astronomy.

American Institute of Physics

One Physics Ellipse

College Park, MD 20740-3843

(301) 209-3100

Web site: http://www.aip.org

The American Institute of Physics is the umbrella organization for many different professional societies of physical scientists. It publishes numerous journals for scientists and magazines for educators, the public, and students interested in careers in physics. Its Center for the History of Physics contains a library and archive of historical books and photographs. The AIP Web site includes a link to that center with a large number of online articles and images, including a discussion of how science came to understand planetary motion.

Jet Propulsion Laboratory (JPL)
California Institute of Technology
4800 Oak Grove Drive
Pasadena, CA 91109
(818) 354-4321
Web site: http://www.jpl.nasa.gov/education/index.cfm
One of NASA's premier research and spaceflight facilities, JPL offers tours, lectures, and events, plus a very informative educational area on its Web site.

The Royal Astronomical Society of Canada
203-4920 Dundas Street W
Toronto, ON M9A 1B7
Canada
(888) 924-7272 (in Canada)
(416) 924-7973
Web site: http://www.rasc.ca
The Royal Astronomical Society of Canada is Canada's leading astronomy organization.

It aims to inspire curiosity in all Canadians about the universe, to share scientific knowledge, and to foster understanding of astronomy for all through activities including education, research, and community outreach activities. Its publications and its extensive Web site have materials for scientists, researchers, teachers, and students of all ages.

The Science Museum
Exhibition Road
South Kensington
London, England SW7 2DD
Web site: http://www.sciencemuseum.org.uk
England's leading science museum, which includes many exhibits related to Newton and features a costumed Isaac Newton character who rides a skateboard as he leads visitors through a twenty-five-minute presentation of Newton's laws and revolutionary ideas.

Smithsonian Institution
Smithsonian Information
P.O. Box 37012
SI Building, Room 153, MRC 010
Washington, DC 20013-7012

(202) 633-1000

Web site: http://www.si.edu

The Smithsonian Institution is one of the USA's most important educational institutions for the public. It consists of many different museums, several of which deal with science and technology.

Woolsthorpe Manor

Water Lane

Grantham, England NG33 5PD

Web site: http://www.nationaltrust.org.uk/woolsthorpe-manor/

Newton's birthplace, including a museum, the original "gravity tree," and a science discovery center.

WEB SITES

Due to the changing nature of Internet links, Rosen Publishing has developed an online list of Web sites related to the subject of this book. This site is updated regularly. Please use this link to access the list:

http://www.rosenlinks.com/RDSP/newt

FOR FURTHER READING

Anderson, Michael. *Pioneers in Astronomy and Space Exploration*. New York, NY: Britannica Educational Publishing, 2013.

Bardi, Jason Socrates. *The Calculus Wars: Newton, Leibniz, and the Greatest Mathematical Clash of All Time*. New York, NY: Basic Books, 2010.

Bortz, Fred. *Physics: Decade by Decade* (Twentieth-Century Science). New York, NY: Facts On File, 2007.

Chiang, Mona. *Isaac Newton and His Laws of Motion.* Pelham, NY: Benchmark Education Co., 2011.

Curley, Robert. *Scientists and Inventors of the Universe*. New York, NY: Britannica Educational Publishing, 2013.

Flood, Raymond, and Robin J. Wilson. *Great Mathematicians*. New York, NY: Rosen Publishing, 2013.

Graham, Ian, David Antram, and David Salariya. *You Wouldn't Want to Be Sir Isaac Newton! A Lonely Life You'd Rather Not Lead*. New York, NY: Franklin Watts, 2013.

Gray, Susan Heinrichs. *Experiments with Motion*. New York, NY: Children's Press, 2012.

Hollihan, Kerrie Logan. *Isaac Newton and Physics for Kids: His Life and Ideas with 21 Activities*. Chicago, IL: Chicago Review Press, 2009.

Lin, Yoming S. *Isaac Newton and Gravity*. New York, NY: Power Kids Press, 2012.

Miller, Ron. *Recentering the Universe: The Radical Theories of Copernicus, Kepler, and Galileo*. Minneapolis, MN: Twenty-First Century Books, 2014.

Robinson, Andrew, ed. *The Scientists: An Epic of Discovery*. New York, NY: Thames and Hudson, 2012.

Ryles, Briony, and Derek Hall. *The Scientific Revolution*. Redding, CT: Brown Bear Books, 2009.

Samuels, Charlie. *Revolutions in Science (1500–1700)*. New York, NY: Gareth Stevens Publishing, 2011.

Timmons, Todd. *Makers of Western Science: The Works and Words of 24 Visionaries from Copernicus to Watson and Crick*. Jefferson, NC: McFarland & Co., 2012.

BIBLIOGRAPHY

Christianson, Gale E. *Isaac Newton and the Scientific Revolution*. New York, NY: Oxford University Press, 1996.

Cropper, William H. *Great Physicists: The Life and Times of Leading Physicists from Galileo to Hawking*. New York, NY: Oxford University Press, 2001.

Encyclopædia Britannica. "Galileo" Retrieved February 19, 2004 (http://search.eb.com/eb/article?eu=108035).

Encyclopædia Britannica. "Sir Isaac Newton." Retrieved February 18, 2004 (http://search.eb.com/eb/article?eu=115657).

Encyclopædia Britannica. "Robert Hooke." Retrieved February 19, 2004 (http://search.eb.com/eb/article?eu=41878).

Giné, Jaume. "On the Origin of the Anomalous Precession of Mercury's Perihelion." Retrieved January 2004 (http://web.udl.es/usuaris/t4088454/ssd/Prepublicaciones/PS/PERIHEL4.PDF).

Gleick, James. *Isaac Newton*. New York, NY: Pantheon Books, 2003.

NASA Glenn Learning Technologies Project. "Newton's Three Laws of Motion." January 2004. Retrieved January 2004 (http://www.grc.nasa.gov/WWW/K-12/airplane/newton.html).

Newton, Isaac. "Hypothesis Explaining the Properties of Light," in Thomas Birch, *The History of the*

Royal Society, Vol. 3 (London: 1757), pp. 247–305. Online at The Newton Project (http://www.newtonproject.sussex.ac.uk/view/texts/normalized/NATP00002).

University of Chicago. "Lecture #9, May 1, 2003." May 1, 2003. Retrieved February 19, 2004 (astro.uchicago.edu/classes/natsci/102/spring-2003/lecture_notes/lecture9.pdf).

University of Oregon. "Dr. Darkmatter Presents The Electronic Universe: Newtonian Physics." January 2004. Retrieved January 2004 (http://zebu.uoregon.edu/~js/glossary/newtonian.html).

The University of Tennessee. "Astronomy 161: The Solar System." January 2004. Retrieved January 2004 (http://csep10.phys.utk.edu/astr161/lect/history/newton3laws.html).

INDEX

S

T

U

W

Y

ABOUT THE AUTHOR

After earning his Ph.D. at Carnegie Mellon University in 1971, physicist Fred Bortz set off on an interesting and varied twenty-five-year career in teaching and research. From 1979 to 1994, he was on staff at Carnegie Mellon, where his work evolved from research to outreach.

After his third book, *Catastrophe! Great Engineering Failure—and Success*, was designated a "Selector's Choice" on the 1996 list of Outstanding Science Trade Books for Children, he decided to spend the rest of his career as a full-time writer. His books, now numbering nearly thirty, have since won awards, including the American Institute of Physics Science Writing Award, and recognition on several best books lists.

Known on the Internet as the smiling, bowtie-wearing "Dr. Fred," he welcomes inquisitive visitors to his Web site at http://www.fredbortz.com.

PHOTO CREDITS

Cover (portrait), pp. 22, 40 The Bridgeman Art Library/Getty Images; cover (apple) © iStockphoto.com/Hans Slegers; p. 5 DEA Picture Library/De Agostini/Getty Images; pp. 8, 10 Photos .com/Thinkstock; p. 13 Three Lions/Hulton Archive/Getty Images; p. 19 Time & Life Pictures/Getty Images; p. 25 Transcendental Graphics/Archive Photos/Getty Images; p. 27 Nicole Russo; p. 31 Stock Montage/Archive Photos/Getty Images; pp. 34, 59 Science Source/Photo Researchers/Getty Images; p. 37 Apic/Hulton Archive/Getty Images; p. 44 Science & Society Picture Library/Getty Images; p. 46 Maciej Oleksy/Shutterstock .com; p. 49 iStockphoto/Thinkstock; p. 53 Photo by Nancy Opitz; p. 57 Digital Vision/Thinkstock; p. 62 NYPL/Science Source/Photo Researchers/Getty Images; cover and interior pages (textured background) © iStockphoto.com/Perry Kroll, (atom illustrations) © iStockphoto.com/suprun.

Designer: Nicole Russo; Photo Researcher: Karen Huang